Meet a Baby Sheep

Jennifer Boothroyd

Lerner Publications • Minneapolis

For the Mona family

Copyright © 2017 by Lerner Publishing Group, Inc.

All rights reserved. International copyright secured. No part of this book may be reproduced, stored in a retrieval system, or transmitted in any form or by any means—electronic, mechanical, photocopying, recording, or otherwise—without the prior written permission of Lerner Publishing Group, Inc., except for the inclusion of brief quotations in an acknowledged review.

Lerner Publications Company
A division of Lerner Publishing Group, Inc.
241 First Avenue North
Minneapolis, MN 55401 USA

For reading levels and more information, look up this title at www.lernerbooks.com.

Library of Congress Cataloging-in-Publication Data

Names: Boothroyd, Jennifer, 1972- author.
Title: Meet a baby sheep / Jennifer Boothroyd.
Description: Minneapolis : Lerner Publications, [2016] | Series: Lightning bolt books. Baby farm animals | Audience: Ages 5-8. | Audience: K to grade 3. | Includes bibliographical references and index.
Identifiers: LCCN 2015038783| ISBN 9781512408003 (lb : alk. paper) | ISBN 9781512412413 (pb : alk. paper) | ISBN 9781512410303 (eb pdf)
Subjects: LCSH: Lambs—Juvenile literature.
Classification: LCC SF376.5 .B66 2016 | DDC 636.3/07—dc23

LC record available at http://lccn.loc.gov/2015038783

Manufactured in the United States of America
1 – BP – 7/15/16

Table of Contents

A Newborn Lamb — 4

The Life of a Lamb — 12

Feeding a Lamb — 17

Growing Up — 24

Why People Raise Sheep — 28

Fun Facts — 29

Glossary — 30

Further Reading — 31

Index — 32

A Newborn Lamb

A mother sheep is cleaning her baby. The baby was just born.

A baby sheep is called a lamb.

When it is clean, the lamb stands up. It takes its first steps.

The lamb is hungry. It drinks milk from its mother.

A mother sheep is called an ewe.

The lamb was born with eight baby teeth on the bottom of its mouth. It doesn't have any teeth on the top.

Ewes can have one to three lambs at a time. Lambs are usually born in the spring.

Lambs are born with short hair called wool. Soon their wool will grow thick and curly.

Lambs get the color of their wool from their parents. The parents pass on genes for wool color. Two gray sheep might pass on gray genes. Or they may pass on genes that make the lamb a different color.

Lambs can be white, black, brown, or gray.

The farmer weighs the lambs a day after they are born. He puts a tag in the lambs' ears to tell them apart.

A newborn lamb weighs only 5 to 8 pounds (2.3 to 3.6 kilograms). Ewes can weigh up to 300 pounds (136 kg)!

The Life of a Lamb

Lambs sleep eight to twelve hours a day. They cuddle with their mother.

Baby lambs jump and run around the pasture. They are playful and curious.

Two lambs play in the pasture.

Lambs bleat to communicate. Ewes can recognize their baby's bleat. A bleat is a *baa* sound.

Have you ever heard a lamb's bleat?

Ewes teach their lambs to follow them around the pasture. The lambs learn how to live with other sheep.

Sheep usually stay together in large groups called flocks. When one sheep starts to move, the rest of the flock follows. Staying with the flock protects sheep from danger.

Feeding a Lamb

For the first few weeks, lambs only drink milk from their mother.

At two weeks, the farmer starts giving the lambs small, healthy food pellets to eat. The pellets are called creep feed.

The farmer puts creep feed into a fenced area of the pen. Only the lambs can fit through the fence to get to the creep.

Lambs can gain about 1 pound (0.5 kg) each day.

Soon the lambs start to graze like their parents. They eat grass, weeds, and hay in the pasture.

Lambs are weaned when they weigh about 50 pounds (23 kg).

The farmer weans the lambs after two or three months. They no longer drink milk from their mother.

This lamb is learning to graze in the pasture.

Sheep chew and swallow their food twice to help with digestion. The food that is chewed and swallowed a second time is called cud.

Sheep also need to drink plenty of water to stay healthy.

Water is at least as important as food for young sheep.

Growing Up

Lambs are called sheep when they are one year old. They begin to lose their baby teeth. Each year, they replace two baby teeth with permanent teeth.

Sheep spend their days grazing, sleeping, and chewing their cud.

Sheep's wool is usually sheared once each year. The wool is cut off and sold or used to make fabric or yarn.

Sheep are able to have their own lambs when they are about one year old.

Sheep can live to be fifteen years old.

Why People Raise Sheep

People raise sheep for many reasons. Wool is the most common product made from sheep. People use wool to make clothes, fabric, carpet, and yarn. They also use wool to produce shampoo, lotion, glue, and ink. Sheep milk is used to make cheeses like feta and ricotta. Many people eat the meat from sheep.

Fun Facts

- Many farmers use other animals to protect their flocks. Dogs, llamas, and donkeys will fight wild animals that want to eat sheep.

- Most farmers raise sheep with white wool. White wool is popular because it can be dyed any color.

- The wool from one sheep is called a fleece. A well-trained shearer can shear a sheep in less than two minutes. The fleece is removed in one large piece.

- Sheep grow 2 to 30 pounds (0.9 to 14 kg) of wool each year. One pound (0.5 kg) of wool makes 10 miles (16 kilometers) of yarn.

Glossary

bleat: a sound a sheep makes

creep: special food for lambs made of easily digested grains like cracked corn

cud: partly digested food that is chewed and swallowed again

digestion: the body's process of breaking down food

flock: a group of the same kind of animal that travel together

gene: information passed from parents to a baby that controls how the baby looks

graze: to eat grass and other plants in a field

lamb: a sheep less than one year old

pasture: a field with plants for animals to graze

pen: an area where an animal lives

shear: to cut the wool off an animal

wean: to get an animal used to eating food other than milk

wool: curly hair of a sheep, goat, or llama, used to make cloth or yarn

Further Reading

Kidcyber: Sheep
http://kidcyber.com.au/tag/facts-about-sheep-farming-for-kids

My American Farm
http://www.myamericanfarm.org/classroom/games

Nelson, Robin. *From Sheep to Sweater.* Minneapolis: Lerner Publications, 2013.

PBS Kids Go: *DragonflyTV*
http://pbskids.org/dragonflytv/show/farmanimals.html

Rudick, Dina. *Barnyard Kids: A Family Guide for Raising Animals.* Beverly, MA: Quarry Books, 2015.

Wilsdon, Christina. *Sheep.* New York: Gareth Stevens, 2011.

LERNER eSOURCE
Expand learning beyond the printed book. Download free, complementary educational resources for this book from our website, www.lerneresource.com.

Index

bleat, 14

creep, 18–19

cud, 22, 25

ewe, 8, 14–15

flock, 16

milk, 6, 17, 21

teeth, 7, 24

wool, 9–10, 26

Photo Acknowledgments

The images in this book are used with the permission of: © iStockphoto.com/rudisill, p. 2; © Margo Harrison/Shutterstock.com, p. 4; © Baloncici/Shutterstock.com, p. 5; © iStockphoto.com/jsorde, p. 6; © iStockphoto.com/pelooyen, p. 7; © J.M. McDonough/Shutterstock.com, p. 8; © iStockphoto.com/levers2007, pp. 9, 31; © iStockphoto.com/MVPixel, p. 10; © iStockphoto.com/JB325, p. 11; USDA, p. 12; © the sea the sea/flickr.com (CC BY 2.0), p. 13; © iStockphoto.com/lauriek, p. 14; © iStockphoto.com/toos, p. 15; © iStockphoto.com/Scott Hortop, p. 16; © iStockphoto.com/Olaf Simon, p. 17; © FLPA/SuperStock, p. 18; © Martin Meehan/Dreamstime.com, p. 19; © patjo/Shutterstock.com, p. 20; © Tambako The Jaguar/flickr.com (CC BY-ND 2.0), p. 21; © Drexie/Shutterstock.com, p. 22; © Johner Images/Alamy, p. 23; © Geothea/Shutterstock.com, p. 24; © iStockphoto.com/Biggunsband, p. 25; © berna namoglu/Shutterstock.com, p. 26; © Photobunnyuk/Dreamstime.com, p. 27; © iStockphoto.com/EricFerguson, p. 28.

Front cover: © iStockphoto.com/pelooyen.

Main body text set in Johann Light 30/36.